Utah
Wildflowers

Utah Wildflowers

A Field Guide to Northern and Central Mountains and Valleys

Richard J. Shaw

UTAH STATE UNIVERSITY PRESS
Logan, Utah
1995

Utah State University Press
Logan, Utah 84322-7800

Book and cover design by Joanne Poon

Cover photographs: front, top, *Primula maguirei*; front, bottom,
Viola nuttallii; back, left, *Aqualegia Formosa*; back, right, *Caltha
leptosepala*

14 12 10 08 06   4 5 6 7 8

LIBRARY OF CONGRESS CATALOGING-IN-PUBLICATION DATA

Shaw, Richard J.
    Utah Wildflowers : a field guide to northern and central moun-
tains and valleys / Richard J. Shaw
        p. cm.
    Includes bibliographic references (p.   ).
    ISBN 0-87421-170-0
    I. Wild flowers—Utah—Identification. 2. Wildflowers—Utah—
Pictorial Works. I. Title.
QK189.S46 1995   94-13838
582.13'09792--dc20   CIP

CONTENTS

Introduction

UTAH WILDFLOWERS is a portable, visual guide to the festival of wildflowers that renew the mountains and valleys of northern and central Utah every spring and summer. The northern two-thirds of Utah contains the Wasatch and Uinta mountains, numerous smaller mountain ranges, and plateaus. Governmental lands encompass six national forests (including eight wilderness areas), one national park, two national monuments, one national historic site, one national recreation area, several protected areas managed by the Bureau of Land Management and the U. S. Fish and Wildlife Service, and several state parks. This book identifies 102 common flowering plants which occur in these areas. Emphasis is placed on naturally occurring wildflowers, but a few alien species have been included because of their abundance and the fact that they are now permanently established. By using the index, common names, and color-keyed photographs, most users of this book will be able to identify and enjoy the flowers of the 21 northern and central counties of Utah (see map inside the front cover).

Why Wildflowers?

When viewing a spreading mountain meadow of glacier lilies mixed with spring beauties, the flower enthusiast may ask the question: "Why do we have so many species and what is the purpose of such delicately arranged petals, sepals, stamens, and ovaries?" Some humans may think that these colorful, sculptured parts are brought together in a flower unit to give us pleasure. Not so. Flowering plants have been evolving these reproductive structures for over a hundred million years, and this long evolutionary history has been linked to the evolution of animals. Today, we have the evidence that biological function lies beneath the delightful surface of these short-lived blossoms. In reality, perpetuation of genes is at stake. What seemingly turns on our human sensations is actually the result of natural selection producing through time an incredible array of precise biological adaptations.

We cannot comprehend the true significance of flowers unless we can visualize something about the coevolution of plants and animals and the development of a spectrum of pollinating agents—insects, birds, and mammals. Some biologists have likened this pollinating relationship to a balanced trading agreement: reward of food in exchange for physically transferring pollen grains from the anthers to the stigmas. The various flower symmetries, the delicate fragrances, and a diversity of startling colors have all appeared because it is effective evolutionary strategy to offer food and color beacons to insects and other animal visitors looking for nectar and high protein pollen. Truly, we all live in highly competitive ecosystems, and the survival of each plant species has been dependent upon developing identifying features which make it stand out from other plants coexisting in the same habitat.

Expanding Your Wildflower Experience

Once the identification of a wildflower has been achieved, greater challenges await the student of these appealing marvels of reproduction. Look for specific characteristics of the whole plant. Is the plant annual, biennial, or perennial? Does it survive in dense shade or full sunshine? Can the soil be described as sandy, gravelly, or loam? Is it possible to recognize a specific plant community—recovery forest, xeric meadow, stream side, or rock outcrop? Observe the flowering data (month, hours of full anthesis). Use a hand lens to observe the surface morphology of the petals, stamens, and pistils. You may be able to determine the method by which the flower is pollinated and the way its fruits and seeds will be disseminated. Your own wildflower garden or your wildflower slide collection will profit from these observations. In any case, this expanded approach will help you learn valuable facts about the intricacies and complexities of these reproductive advertising agents—the wildflowers.

Plant Names

In this book each entry is headed by the plant's common name or names. These are taken from identification manuals of flowers growing in the Intermountain West. It should be understood, however, that common names are not universally accepted and as yet are not applied with any set of accepted rules. The botanical name, in italics, appears below the common name and is based on international rules which use a standardized two-part naming system called binomial nomenclature. The first part of the botanical name is the generic name and the second part of the name is the specific epithet. Thus every plant species has two names.

This binomial will be accepted around the world and cannot be changed unless certain international rules of nomenclature are followed.

Botanical names have Latin endings and are easier to pronounce than they may appear to be. Generally, you simply say the word as you would any English word. Regardless of how many syllables the word has, pronounce each syllable as you would in any ordinary word, slowly and distinctly. Don't worry about stressing any one syllable.

Example:

Heracleum lanatum HER-AK-lee-um LAN-a-tum

Mentzelia laevicaulis MENT-zee-lee-a LEVEE-caul-is

Photographing Wildflowers

With the photographic equipment and film that are available today the vast majority of wildflowers can be photographed with a minimum of patience and effort and with an appreciation of the subject. Sunshine is best but some of the most beautiful pictures of wildflowers have been taken in the shade or even in the rain by using flash attachments. For flower close-ups of 6 to 18 inches, a macro lens is worth the expense in terms of depth of field and focusing ease. A tripod is essential because one can compose each picture for maximum detail and appropriate background, and if a breeze has the flower stems moving, one can wait for a momentary lull to press the shutter.

Backgrounds may be provided by natural shadows or even blue sky if a low shooting angle can be found. Natural objects in the plant's environment such as a downed tree may be helpful to give a feeling of a wilderness setting. However, there may be times when you want to emphasize flower structure with maximum contrast. This is the time to use an artificial background. A piece of black velvet or paper held behind the flower may be all that is needed. Film is rather inexpensive in terms of potential results, so take several shots of the flower subject using different angles and exposure times.

Finally, do not be guilty of micro-habitat destruction. Trampling a hundred flowers to photograph one flower is as detrimental to a habitat as a bull in a china closet. Be careful where you place your feet and your knees. Such patience and effort will be rewarded as you view slides and prints years later.

Organization of Flowers by Color

To aid rapid recognition, the photographs are arranged into five color groups: (1) green to cream-colored and white, (2) yellow to orange, (3) pink to red, (4) blue to purple, and (5) brown to reddish brown. Because each person has slightly different perceptions of color, such a scheme of identification has its limitations, especially in the pink to red and in the blue to purple categories. Be sure to check another likely color if there is any question. Also, remember that some species have intergrading specimens when it comes to petal color.

Wildflowers for the Future

Picking wildflowers or collecting specimens of animals, trees, minerals, or archaeological artifacts in all National Parks and Monuments is prohibited without special permission from the park superintendent. The same policy applies to other governmental agencies. Study the plants where they grow, take photographs of them home with you, but leave them for the enjoyment of those who will follow.

Wildflowers by Color

Tufted Evening Primrose

Oenothera caespitosa
Evening Primrose Family

This plant, also known as moonrose, is found on open sunny slopes during April, May, and June. Little or no stem is visible above the tufted rosette of leaves. Like other members of the family, the flowers are composed of four sepals, four petals, eight stamens, and, in this species, four diverging stigmas. The sweet scented blossoms are 2 to 2½ inches broad and each petal is deeply notched at the tip. When the flowers first open they are white, but they turn pink as they mature in the next 24 hours. Nectar glands are situated at the base of the floral tube, which may be up to 3 inches long. Only those insects with long mouth parts, such as hawk moths, are capable of reaching the nectar. The ovary remains hidden among the leaf bases and matures into a many-seeded capsule.

Oenothera caespitosa

Common Cow Parsnip

Heracleum lanatum
Carrot Family

This pubescent perennial herb is 3 to 8 feet tall and features flower clusters 4 to 6 inches broad. It is found in moist sites, particularly in the canyons. The base of each leaf forms a conspicuous sheath around the coarse stem. The umbel is compound and consists of 15 to 30 umbellets, and the outer flowers of each umbel have larger petals. Native Americans made good use of this plant in various ways—from using them in the Sun Dance ceremony to using a strong mixture of the roots in the treatment of rheumatism and arthritis. To render the hollow stems palatable to humans, they can be peeled and boiled twice in clean water. However, no member of the carrot family should be eaten unless positive identification is possible.

Heracleum lanatum

Sego Lily

Calochortus nuttallii
Lily Family

The generic name *Calochortus* comes from two Greek words meaning "beautiful grass," and this flower has long been recognized as one of the most beautiful members of the lily family. Three narrow sepals provide a sharp contrast to the three broad, conspicuously marked petals. The base of each petal has a round, yellow gland surrounded by bright yellow hairs and a purple crescent marking outward from the gland. The bulbs, which are 5 to 6 inches below the surface, are sweet and nutritious and can be eaten cooked or raw. Native Americans and early western settlers relied on them when other food was scarce. Several *Calochortus* species once thrived in the western states, but cities and towns encroaching upon foothill habitats have left some species in a precarious position. This state flower of Utah blooms during June and July.

Calochortus nuttallii

13

Colorado Columbine

Aquilegia coerulea
Buttercup Family

Aquilegia comes from the Latin word *aquila* meaning "eagle," referring to the long, spurred petals which look like eagle talons. Throughout July and August this delicate and ornamental wildflower brightens canyon trails up to 9,500 feet. The plant has one or more stems 4 to 16 inches tall and mostly basal leaves that are three parted and lobed. The sepals may vary in color from blue to white. Numerous yellow stamens and five long pistils project beyond the flower face. The species in this genus have few barriers to gene exchange and will hybridize readily. Specific pollinating agents tend to keep the species isolated. This columbine is the state flower of Colorado.

Aquilegia coerulea

Douglas Chaenactis

Chaenactis douglasii
Sunflower Family

This plant has inspired a number of vernacular names, including brides bouquet, dusty maiden, false yarrow, and hoary chaenactis. Such names reflect features of the composite head and foliage. A basal rosette of leaves appears first, and these are followed by leafy stems 12 to 18 inches tall. The individual leaves are fernlike because they are pinnately dissected. Ray flowers are lacking, and only white to pink tubular disk flowers are present. The Paiute Indians crushed the leaves and applied them as a poultice. Dry, gravelly roadside cuts are the best sites to watch for this species from June through September. An alpine species of this genus, *C. alpina*, has a more pink disk flower and a small stature of about 3 inches.

Chaenactis douglasii

Leafy Jacobs Ladder

Polemonium foliosissimum
Phlox Family

      This tall member of the phlox family often reaches a height of 3 feet and will be found in canyons along streams or small lakes. The five sepals have glandular hairs, the five white petals flare gradually from their united base, and the five large anthers have conspicuous yellow pollen. The petal color may vary from white to blue, but in northern Utah the color is predominantly white. The leaves are pinnately lobed with up to 25 lobes. Flowering occurs from June to August, and sometimes this species is confused with *P. occidentale*, western Jacobs ladder, but the latter has only a solitary stem.

Polemonium foliosissimum

19

Prickly Poppy

Argemone munita
Poppy Family

The generic name is from the Latin *argemon* meaning "cataract," which the plant was supposed to cure. At first glance, a prickly poppy along a roadside might appear to be a thistle with numerous prickly leaves, but a closer look at the flower reveals four to six delicately crumpled petals with a ball-shaped mass of stamens in the center. As with most poppies, the flower buds have two or three sepals equipped with a stout horn. The leaves and stems are covered with spines and prickles, which discourage livestock. In addition the latex-laden sap is distasteful. The toxic seeds are numerous and contained in a cone-shaped capsule that eventually splits open to release numerous black seeds. Small black beetles are often seen crawling on the stamens and petals. Flowering occurs from June through August.

Argemone munita

21

Marsh Marigold; Elkslip

Caltha leptosepala
Buttercup Family

For the high mountain hiker, the thrill of this plant comes with the discovery that blue-colored buds push through the melting snow. Within 48 hours, these blue buds have expanded into sparkling white blossoms, 1 to 2 inches in diameter, that are similar to buttercups and anemones. The flowers of this genus lack petals, but the sepals are petaloid and showy. The specific epithet *leptosepala* means "thin" or "narrow," referring to the sepals. Each flower has many stamens and five or more pistils. The basal leaves are heart shaped and slightly scalloped on the edges. Some authors have recommended this plant as an edible pot herb; however, the leaves contain helleborin which has a burning taste and is also toxic.

Caltha leptosepala

Arctic Gentian

Gentiana algida
Gentian Family

Visit the high Uinta Mountains during late summer and look for one of our most esteemed wildflowers. *Algida* means "cold," referring to the severe climate of the alpine ecosystem. In gentians the calyx of the flower is a cup or tube with four or five teeth; the corolla also forms a tube or funnel with four or five pleated lobes. Arctic Gentian is easily recognized by the delicate blue to purple specks or splotches which decorate both sides of the petals. The basal leaves of this perennial are long and narrow with smooth margins. The gentian family has over 1,000 species, and 11 of these occur in Utah.

Gentiana algida

False Hellebore; Skunk Cabbage

Veratrum californicum
Lily Family

In canyons and subalpine meadows, this plant is a tall broad-leaved herb reaching up to 5 or 6 feet high. The leaves are strikingly large, alternate, pleated, and parallel veined. The typical lily-like flowers are white to greenish, are numerous, and grow in large terminal clusters. All false hellebores contain alkaloids and several of these have been used in medicines and insecticides. Eating this plant can cause depressed heart activity, headaches, and general paralysis. Consumption of these plants by female sheep the first three weeks after pregnancy can cause abortion or birth defects. The generic name *Veratrum* means "truly black," referring to the roots of *Veratrum album*.

Veratrum californicum

Woodland Strawberry

Fragaria vesca
Rose Family

Hiking along trails of the open woods in a mid-montane environment, many hikers will immediately recognize the familiar fruits and flowers. Even though this plant lacks erect stems, it produces horizontal stolons or runners which, in turn, form new plants at their tips. Such vegetative reproduction is very efficient. Each flower consists of five green sepals, five rounded petals, and many stamens. The pistils are scattered on a conical hump of tissue which becomes part of the edible fruit. Herbalists of the last four centuries have had a high regard for the strawberry's medicinal properties. For a tea high in vitamin C, steep 2 handfuls of leaves in 1 quart of boiling water for 5 minutes.

Fragaria vesca

Richardson Geranium

Geranium richardsonii
Geranium Family

The generic name *Geranium* means "crane flower" and refers to the long-beaked fruit, which looks like a crane's bill. Like its close relative, *G. viscosissimum,* Richardson geranium begins to bloom in June and continues into late August. However, its habitat requirement is quite different. The Richardson geranium grows in moist aspen-fir woodlands, often by a small stream, instead of in the sagebrush community or open woods. The petals are about $1/2$ inch long and are streaked with pinkish to purple guide lines. The five parts of the ovary separate, and the seeds are catapulted outward by the drying segments of the style. The leaves are palmately lobed into several pointed segments.

Geranium richardsonii

Engelmann Aster

Aster engelmannii
Sunflower Family

In the canyons above 6,000 feet, especially along streams and meadows, the Engelmann aster may grow to a height of 3 feet. Making an identification between members of the genus *Aster* and the genus *Erigeron* (daisies) is not easy. Many technical features need to be examined. However, this species is easily recognized by such reliable criteria as a yellow disk flower surrounded by 15 to 20 white ray flowers that may be 1 inch long. The leaves are lanceolate to elliptic and nearly smooth. The greenish bracts, which enclose the composite head of flowers in bud, form a structure known as an involucre, and, therefore, each bract is called an involucral bract. Flowering begins in July and continues into September.

Aster engelmannii

Fendler Meadowrue

Thalictrum fendleri
Buttercup Family

The generic name *Thalictrum* comes from the Greek language meaning "the blooming one," and species in the genus have a long list of herbal uses. Meadowrue's fern-like foliage is often confused with the related columbines. The branched, sometimes purplish stems can reach a height of 2 feet. This species is dioecious, meaning that one plant will bear only male flowers and another different plant will bear only female flowers. These unisexual flowers are unique in that they have five inconspicuous green sepals, but no petals. The downward hanging male flowers are most apt to catch one's eye because the numerous, greenish anthers dangle from fragile filaments. The whole flower has the appearance of a lamp shade with a fringe on the bottom. The female flowers bear 6 to 10 pistils. Flowering occurs in June and July.

Thalictrum fendleri

Green Gentian; Monument Plant

Frasera speciosa
Gentian Family

On open slopes or in open forest, the robust green gentian is one of the tallest of the herbaceous plants. It sends forth a single stem 3 to 5 feet high. Leaves 10 to 20 inches long form evenly spaced whorls on the stem. Each individual flower has a plan of four narrow sepals, four broad petals, and four stamens. Many insects are attracted to the elaborate nectar glands and purple spots on each petal. The fruits are large capsules with a diameter of $1/2$ inch and produce copious amounts of small seeds. While the plant is perennial, it flowers only once and then dies. In Utah it ranges from 5,500 to 10,000 feet and blooms from June to August. The specific epithet *speciosa* means "showy" or "beautiful" and refers to the flowers.

Frasera speciosa

Canada Violet

Viola canadensis
Violet Family

Northern and central Utah have at least nine species of naturally occurring violets. Recognition of this species is relatively easy because of its white flowers and its leaf blades that are often wider than long and broadly heart shaped. Even though the petals appear snow white, the throat of the corolla changes abruptly to lemon yellow. The lower petals of this bilaterally symmetrical flower have several distinctive purple lines that are called guide lines by students of pollination. There is good evidence that these lines and the hairs in the throat direct insect visitors past the stamens to the nectar in the basal spur. This perennial blooms in the canyons and montane forests from June through August. Elias and Dykeman suggest that all of the species of violets are edible either raw in salads or cooked as pot herbs.

Viola canadensis

Fringed Grass-of-Parnassus

Parnassia fimbriata
Saxifrage Family

This beautiful inhabitant of springs, streams, and lake shores has heart-shaped to kidney-shaped leaves, and therefore, the common name, grass-of-parnassus, is misleading. The solitary flowers are on the ends of stems 5 to 12 inches high. The photograph highlights the daintily fringed petals which are so characteristic of this species. Note also that there are five white fertile stamens alternating with five yellow sterile stamens which look like nectaries. The generic name refers to the mountain in Greece where the muses of mythology lived, and the specific epithet *fimbriata* means "fringed." July and August are prime months to see the blossoms.

Parnassia fimbriata

Common Yarrow

Achillea millefolium
Sunflower Family

The generic name *Achillea* is named for the Greek hero, Achilles, who is said to have used yarrow to heal the wounds of his soldiers. The specific epithet *millefolium*, "of a thousand leaves," describes the fern-like dissection of each alternate leaf. The composite flowering head has both ray and disk flowers, and the corolla color of the ray flowers is mostly white and occasionally pink. The plant is known throughout the northern hemisphere, from sea level to the alpine zone. With such genetic plasticity, it is easy to see why several subspecies have been proposed. From the time of the ancient Greeks, yarrow has kept its place among medicinal plants, being used for toothaches, urinary problems, and head colds.

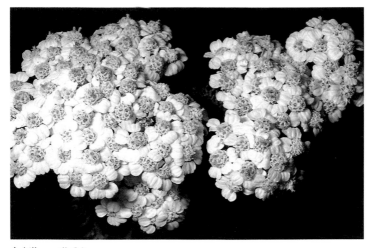

Achillea millefolium

43

Small-flowered Woodlandstar

Lithophragma parviflorum
Saxifrage Family

Woodlandstars are among the first plants to flower in the spring and live in foothills and canyons and on exposed rocky slopes. Most of its leaves, which are lobed, are at the base of the stem. The flowers are in racemes with white or pink petals over $^{1}/_{4}$ inch long. Note how each petal is deeply cleft into two or four narrow lobes, a feature that ties in with the common name. Within the calyx and corolla are 10 stamens. The generic name comes from the Greek words lithos "stone," and *phragma* "rock," apparently referring to the plant's habitat. Within Utah there are two other species of the genus, one of which, *L. glabrum,* forms bulblets in the inflorescence and axils of the stem leaves.

Lithophragma parviflorum

45

Water Buttercup; Water Crowfoot

Ranunculus aquatilis
Buttercup Family

The generic name *Ranunculus* comes from the Latin *rana*, meaning "frog," and refers to the wet habitats of some species. The stems of this species float under and upon the surface of shallow ponds, lakes, and streams, especially in the canyons. During July and August, these plants can entirely cover the slow moving water. The five white petals are about $1/2$ inch long and may be yellow at the base. Numerous stamens and pistils complete the flowers which are usually lifted above the water by slender pedicels. This truly aquatic plant has two types of leaves. The floating leaves are deeply cleft with three to five lobes, while the submerged ones are dissected into forked, hair-like segments.

Ranunculus aquatilis

Wartberry Fairybell

Disporum trachycarpum
Lily Family

In the canyons under the shade of maples, oaks, and low shrubs, this branched perennial with undulating, veiny leaves often goes unnoticed. The paired narrowly bell-shaped flowers, about $1/2$ inch long, are often hidden as they hang beneath the leaves at the end of pedicels. By July, the flowers give rise to berries with wart-like projections. As the berries ripen they change from yellow to orange to deep red. While some liliaceous plants produce poisonous fruits, these berries are edible either cooked or raw. Flowering begins in May.

Disporum trachycarpum

American Bistort

Polygonum bistortoides
Buckwheat Family

Polygonum comes from the Greek meaning "many knees," referring to the swollen stem nodes. The upper leaves are narrow and tapering. The flower stem is 12 to 20 inches high. The individual flowers are small and bear five petaloid parts with exserted stamens. The rhizomes of this species have often been used by Native Americans, who prize them highly for their starchy and rather pleasant taste. The name, bistort, comes from the Latin words meaning "twice twisted," referring to the gnarled appearance of the dark brown rhizomes. This frequent herb of subalpine meadows and stream banks blooms during July and August.

Polygonum bistortoides

51

Foothill Death Camas

Zigadenus paniculatus
Lily Family

This death camas is among the most infamous of the western plants, because it has poisoned livestock, especially sheep, and occasionally humans who have mistaken the egg-shaped bulbs for the edible bulbs of common camas. These perennials resemble wild onions but lack the onion odor. Leaves are long, slender, and diverging upward. The flowers are usually about $1/8$ inch in diameter with three sepals, three petals, and six stamens. All parts of the plant contain alkaloids that can cause gastrointestinal problems, weakness, loss of motor function, and death. Two other species of *Zigadenus* in Utah are considered poisonous. May and June are flowering periods, particularly in the foothills of the Wasatch Mountains.

Zigadenus paniculatus

53

False Solomon's Seal

Smilacina racemosa (Maiantheum racemosa)
Lily Family

This plant is called "false" since it has none of the medicinal properties of the true solomon's seal or spikenard (genus *Polygonatum*). From a horizontal rhizome, erect leafy stems grow up to 18 to 20 inches high. The veiny leaves are 3 to 6 inches long and clasp stems at the base. In spite of the specific epithet, *racemosa,* the dense cluster of flowers occur in panicles. The six stamens are slightly longer than the six petaloid segments and have a wide, flattened filament. The berries are speckled in the maturation stage, but eventually turn a bright red. Some authors suggest that the red berries are edible but have a purgative effect. Look for the blossoms in May and June in the woodland shade.

Smilacina racemosa
(Maiantheum racemosa)

Showy Goldeneye

Viguiera multiflora
Sunflower Family

Long after the spring and early summer flowers have faded from the foothills and canyons, many roadsides and mountain trails are brightened by the golden-yellow heads of this yellow composite. This herb grows up to 3 feet tall, with few to several heads per stem. The leaves are lance shaped and oppositely arranged, at least in the lowermost part. The strikingly bright ray flowers are $\frac{1}{2}$ to 1 inch long. The numerous disk flowers are accompanied by scales called chaff, which partially envelop the fruits (achenes). The disk flowers are also somewhat darker, suggesting a golden eye. There is growing evidence that this species is responsible for the poisoning of grazing animals in the southwestern states.

Viguiera multiflora

Glacier Lily

Erythronium grandiflorum
Lily Family

Common names of plants frequently cause confusion among flower lovers because they vary so much in geographical areas. For example, this lily is also known as dogtooth violet, adders tongue, fawn lily, and trout lily. In northern and central Utah, the plants are often locally abundant above 7,500 feet at the edge of melting snowbanks. The yellow nodding flowers are on stems 6 to 12 inches high. The six tepals (sepals and petals) curve backward exposing six stamen and three stigmas to full view. Anther color varies from red to yellow. The shiny, broad leaves and flowering stalk emerge from an underground bulb which is readily eaten by bears. Flowering is from May to July, depending on how rapidly the snow fields melt.

Erythronium grandiflorum

Butter-and-Eggs; Common Toadflax

Linaria vulgaris
Figwort Family

In his folklore of North American wildflowers, Coffey lists over 27 vernacular names for this European native which was a weed in flax fields. In North America the plant has often escaped cultivation and formed large populations from creeping rhizomes. It closely resembles the cultivated snapdragon, except that the inch-long corolla is spurred at the base. Slender stems, 1 or 2 feet tall and bearing numerous narrowly linear leaves, arise from perennial roots. The name toadflax refers to the corolla's "mouth" resembling that of a toad and to leaves which look like flax leaves. Since the plant grows in disturbed sites from the valleys to subalpine, it can be in flower from June through August.

Linaria vulgaris

Blazing Star Mentzelia

Mentzelia laevicaulis
Blazing Star Family

Occupying gravelly roadside cuts and dry stream beds, the blazing star lives up to its common name. The light yellow flowers, which are up to 1 to 3 inches in diameter, are borne in a branching inflorescence at the ends of 2 to 3 foot stems. The delicate, numerous stamens are nearly as long as the petals and fan outwards in all directions. Because the barb-like hairs on the leaves and stem adhere to the hair of animals and the clothing of humans, these plants are sometimes called stickleleaf. Native Americans of Montana dug the roots before the plants flowered and found them useful for fevers and complex illnesses. The flowers give rise to cylindrical woody capsules.

Mentzelia laevicaulis

Nuttall Violet

Viola nuttallii
Violet Family

Northern and central Utah have at least two species of yellow violets which bloom during May and the first two weeks of June. Look for them in sagebrush and mountain brush communities. The one pictured here is the larger of the two and its larger leaf is up to $2\frac{1}{2}$ inches long. All violets have bilaterally symmetrical flowers, consisting of five sepals, five petals, and five stamens. The lowest petal bears a sac-like spur at its base and contains nectar. The flower structure, including the brownish to purple guide lines, favors cross pollination. The genus *Viola* is considered a critical group with many difficult to recognize species (with at least 10 species in northern Utah). Hybridization occurs freely under natural conditions, and, in some cases, the species of *Viola* produce inconspicuous flowers and seeds hidden at the base of the plant.

Viola nuttallii

65

Mountain Goldenpea

Thermopsis rhombifolia
Pea Family

Because of the superficial resemblance, this plant is frequently called false lupine, but several distinctive features set it apart. The leaves have only three leaflets while lupines have five or more. The 10 stamens are always distinct, and lupines and milkvetches have several stamens fused. As the petals fall off, the ovary elongates into a flat pea-like pod. Goldenpea is unpalatable to wildlife and grazing livestock, so it may replace more desirable forage plants. There are many reports that this legume contains a number of toxic alkaloids, especially in the seeds. This species has surprising genetic plasticity and can tolerate different habitats, including wet meadows in valleys and dry midmontane forest sites. Flowering begins in May and continues through July.

Thermopsis rhombifolia

Arrow Leaf; Balsamroot

Balsamorhiza sagittata
Sunflower Family

Northern Utah has three species of balsamroots, but the one pictured here is the most common. In May and early June, foothills and open forest slopes come alive with a carpet of yellow flowers. The numerous basal leaves up to 1 foot in length are arrowhead shaped and covered with tiny, silvery hair on both sides. The stems are 1 to 2 feet tall, bearing solitary heads at their apex. Like many other sunflower-like plants, the conspicuous ray flowers are pistillate only, while the disk flowers are always perfect. The odoriferous, deeply penetrating roots and ripe achenes of this plant were used by Native Americans as a source of food. Where two or more species of *Balsamorhiza* grow together they may form intermediate hybrids.

Balsamorhiza sagittata

Mules-ear Wyethia

Wyethia amplexicaulis
Sunflower Family

While superficially looking like balsamroot species and often occupying the same habitats, the mules-ear can be easily separated from other sunflowers. The leaves are the first feature to examine, being long and hairless and appearing to be varnished. They also clasp the stem, and, hence, the specific epithet is appropriate—*amplex* meaning "embrace" and *caulis* meaning "stem." The second feature to examine lies with the flowering heads. The plant will produce two or more heads per stem. Goshute Indians ate the seeds and applied the roots externally as a remedy for bruises and swollen limbs. The generic name *Wyethia* is in honor of Captain N. J. Wyeth who crossed the continent with an early botanist in 1834. Late May and early June bring forth the flowers.

Wyethia amplexicaulis

Western Wallflower

Erysimum asperum
Mustard Family

Within the northern and central counties of Utah, there are many wild mustards with white and yellow flowers, but this is the only species with flowers over $\frac{1}{2}$ inch long. The flowers have four sepals, four petals at least $\frac{1}{2}$ inch long, six stamens in two sets (four long and two shorter), and distinctive elongated pod-like fruits reaching a length of 2 to 3 inches. The one or more stems may reach 24 inches in length, and bear leaves which are narrow and sometimes have margins with small teeth. The startling beauty of this plant should be looked for in late May and early June. Zuni Indians ground this plant, mixed it with water, and applied it to their skin to prevent sunburn. The common name comes from related species that grew on ancient rock walls of Europe.

Erysimum asperum

Orange Agoseris; Mountain Dandelion

Agoseris aurantiaca
Sunflower Family

Usually found in the mountains above 6,000 feet, this stunning orange dandelion will grow among grasses in subalpine meadows. It is a perennial plant with a woody taproot and one to several leafless stems up to 18 inches high. When broken or cut, the stems and basal leaves exude a milky juice. All the flowers in the composite head are ray flowers, which change to brownish-purple as they age. The fruits that follow have soft, white hairs which greatly aid in dispersal. When fully mature, each fruit is like a miniature parasol. Like the dandelion in our lawns, these plants can be used as potherbs and for making tea and wine. Flowering will continue from June to August.

Agoseris aurantiaca

75

Yellow Monkeyflower

Mimulus guttatus
Figwort Family

Fragile and appealing beauty characterizes this variable herb of wet meadows and stream banks. After recognizing that the flower has bilateral symmetry, one can see that delicate hairs cover the three lower lobes of the corolla, and, together with the reddish-orange spots, these hairs help to attract insect pollinators. Close examination of the stigma reveals two roundish lobes which are spread apart. When one of these lobes makes contact with a pollen-laden bee, the two stigma lobes immediately come together like the covers of a book. Thus the pollen will be held firmly, and when the bee backs out of the corolla tube, no self-pollination will occur. This species, which blooms from June through August, is one of at least eight species of the genus in Utah. Height of the plant can vary from 4 to 20 inches.

Mimulus guttatus

Yellowbell; Yellow Fritillary

Fritillaria pudica
Lily Family

The bleakness of winter fades rapidly once the yellowbells pops through in late April and all of May, depending on elevation. The yellow or orange drooping flowers turn reddish with age. The perianth parts are ¾ inch long, and when they fall and the fruit starts to ripen, the stems straighten out, placing the three-sectioned fruit in an erect position. The underground bulbs are 3 to 6 inches below the surface and reproduce asexually by forming tiny bulblets the size of corn kernels. The narrow leaves are surprisingly succulent and grow in pairs or whorls of three or more. While the bulbs and bulblets were eaten by Native Americans, the use of such food should be discouraged, as sagebrush habitats are fast disappearing with the encroachments of towns and cities.

Fritillaria pudica

Heartleaf Arnica

Arnica cordifolia
Sunflower Family

This low perennial is widely distributed in the Utah canyons, especially under aspen and spruce-fir forests. It receives its common name from the fact that the basal leaves and the opposite leaves of the stem are markedly heart shaped at the base. The plants are mostly 8 to 13 inches tall, and the flowering heads are 2 to 3 inches in diameter. The leaves have a sawtooth margin, and the plant spreads from extensive rhizomes. Various species of *Arnica* have been used medicinally in Europe since the sixteenth century, but this species has been listed in the Southwest as a poisonous plant because of the presence of arnicin, a crystalline toxin, in the leaves. Blossoming begins in late May and lasts to mid-July.

Arnica cordifolia

81

Canada Goldenrod

Solidago canadensis
Sunflower Family

In the herbals of the past, goldenrods were described as having healing properties. In fact, the generic name means "make whole." It is frustrating to separate the many species of *Solidago* found in North America because the genus seems to lack consistent distinguishing characteristics. The species in Utah are all erect perennials, bearing alternate leaves and small composite heads which contain both ray and disk flowers. The numerous, small flowering heads are arranged on only one side of the spreading branches. This tall species, up to 4 feet in height, grows in a variety of dry, open sites and flowers in late summer. Goldenrods were once considered a major cause of hay fever, but such beliefs have been discredited by studies showing the flower to be pollinated by bees. Because the pollen is relatively heavy, it is poorly dispersed by the wind.

Solidago canadensis

Yellow Pondlily

Nuphar polysepalum
Waterlily Family

Within the beaver ponds and small lakes of the Uinta Mountains, this large aquatic plant grows from submerged rhizomes and bears long petioled leaves which float on the surface of the water. Its attractive flowers appear in July and August. They are 2 to 3 inches in diameter, having five to six sepals and many scale-like petals. The stamens are numerous and close to the flattened stigma. The glossy leathery leaves are 6 to 15 inches long and 6 to 10 inches broad. The rhizomes are buried in the mud and were often gathered by Native Americans, who, in Montana, parched the seeds or ground them into flour.

Nuphar polysepalum

King Flax

Linum kingii
Flax Family

Appearing on exposed rocky slopes of the subalpine during July and August, this infrequently seen flax species is remarkably like its blue relative, wild blue flax. The flowers are radially symmetrical, with five separate sepals, five separate petals which easily drop off if disturbed, and ten stamens, joined by the bases of their filaments. The fruits are five-chambered capsules. The generic name *Linum* comes from the Greek *linon*, meaning "flax." All wild species of flax have stems which produce strong fibers. Native Americans have used these fibers effectively in making baskets, mats, and even the mesh of snowshoes.

Linum kingii

Straightbeak Buttercup

Ranunculus orthorhynchus
Buttercup Family

Northern Utah has at least twenty species of buttercups, ranging from the valleys to the alpine. These species are not easily separated, and a plant taxonomist may be needed for proper identification. Straightbeak buttercup typifies most of the mountain species. Flowers usually have five sepals and petals surrounding numerous stamens and pistils. The generic name *Ranunculus* is derived from the Latin word *rana*, meaning "frog." The name is suggestive of moist habitats where most species are found. The fruits and young plants contain an irritant alkaloid which, if taken internally, may cause nausea or vomiting. Contact with the leaves and sap may cause dermatitis in susceptible persons. The plants lose their toxicity when dried.

Ranunculus orthorhynchus

Summer Pheasant-eye

Adonis aestivalis
Buttercup Family

Pheasant-eye is among the European aliens which have become permanently established in Utah's valleys, fields, and roadsides. While it is a poisonous weed that threatens horses and sheep, few people could say it is not a beautiful eye catcher when it flowers in May and June. The erect stems reach 5 to 24 inches tall and bear alternate, pinnately dissected leaves. The five to eight petals are separate and elliptical and usually have a black spot at the base. Stamens are more than 10 in number, and their black anthers contrast well against the red-orange perianth. After the petals fall off, an elongate cluster of about 50 achenes begins developing. Here is an annual which is well suited to invade recently disturbed sites in low lying valleys.

Adonis aestivalis

Munro Globemallow

Sphaeralcea munroana
Mallow Family

There are several species of globemallows on western ranges associated with sagebrush and dry foothill slopes. The orange-flowered species pictured here can be recognized by the lobed-only leaves rather than the palmately divided leaves of some of the other species. Using a hand lens on the leaf surfaces will reveal some delicate stellate hairs. Like other mallows, the 1½ inch flowers have five petals, five pubescent sepals, and many stamens, all united at the base. When immature, fruits are roundish, and this feature accounts for the generic name (*Sphaer* "sphere," and *alcea* "mallow"). The mature fruits split open at maturity, resulting in pie-shaped segments, each with a single seed. This many-branched perennial herb may start blooming in May and continue into mid-summer.

Sphaeralcea munroana

93

Oregon Grape; Creeping Barberry

Mahonia repens
Barberry Family

In the canyons and foothills, and under a variety of woody plants, this low-growing shrub sparks the spring season with dense clusters of yellow flowers that brighten the appearance of prickly evergreen leaves. Even though there is no clear distinction between the sepals and petals, there are usually six sepals, six petals, and six stamens. By August the plants have produced rather sparse clusters of bluish-purple grape-like fruits. Some people claim the berries have a bitter taste, but a number of authors recommend using them in making jelly, jam, or wine. Along with the ripening of the fruit, some of the pinnately compound leaves turn beautiful shades of red or purple. This species transplants well into a native garden with some shade.

Mahonia repens

95

Lanceleaved Stonecrop

Sedum lanceolatum
Stonecrop Family

Lanceleaved stonecrop flowers from late June through August and represents a family of succulent plants which have water laden stems and leaves. Recognizing plants of the genus *Sedum* is easy, but separating the species is sometimes confusing, and calls for special attention to details. This particular stonecrop has numerous basal leaves. While the leaves of the stem vary greatly in shape, they are not ridged underneath. Each flower, resembling a bright yellow star, has five petals, eight to ten stamens, and five pistils which form five follicles when mature. Look for this species on rocks and gravelly soil, extending from sagebrush-covered foothills upward to the subalpine. The young stems and fleshy leaves are reported to be good eating when cooked.

Sedum lanceolatum

6-2—12
Grand Canyon

Alpine Sunflower; Rydbergia

Hymenoxys grandiflora
Sunflower Family

The almost perfect symmetry of the ray and disk flowers of this alpine species are in contrast to the rocky exposed ridges where it is found. The flowering head is 2 to 3 inches in diameter, and each ray flower has three lobes at the tip. Another common name of this plant, old-man-of-the-mountain, refers to the dense hairs on the stems and the linear lobes of the leaves. This pubescence reduces water loss, traps heat, and protects against ultra-violet radiation. The flowering heads usually face east, perhaps for protection from the prevailing wind. Zwinger and Willard say that each plant stores food over several years until it has sufficient nutrients and energy to blossom, and after the fruits and seeds mature the entire plant dies.

Hymenoxys grandiflora

Puccoon; Western Stoneseed

Lithospermum ruderale
Borage Family

The generic name comes from the Greek *lithos* meaning "stone," and *sperma* meaning "seed." The name refers to the stony nutlets which replace the lemon-yellow flowers. During May and June the flowers seem to be partially hidden among the numerous, narrow leaves of the stem. The five petals are fused into a narrow tube, and the five stamens are attached to the inside of the united corolla. Puccoon occupies a number of habitats in Utah, from sagebrush foothills to dry sites in the midmontane. Shoshoni women of Nevada reportedly drank cold water infusions of stoneseed root daily as a contraceptive. Tests with mice have shown that extracts of this plant eliminate the estrous cycle (Weiner 1972).

Lithospermum ruderale

Hairy Golden Aster

Heterotheca villosa
Sunflower Family

Species of this genus resemble the genus *Aster* in a number of ways, but differ in that the ray flowers are yellow instead of being white to purple. One feature, visible under a hand lens, is the presence of two whorls of bristles on scales on the numerous disc flowers. Stems of the species illustrated grow 6 to 20 inches tall, bloom mostly in August, and are common on open, rocky slopes from the foothills to the canyons. *Villosa* means "hairy," referring to the soft pubescence which covers the stems and leaves. This covering is not sticky as it is in gumweed. This is a complex and confusing species which consists of several varieties differing in such details as pubescence.

Heterotheca villosa

Common Mullein

Verbascum thapsus
Figwort Family

Here is a Eurasian weed which has virtually spread throughout North America since colonial times. It has the distinction of having at least twenty common names, such as miners candle and flannel-weed. This biennial plant of roadsides forms a basal rosette of leaves the first year, with the leaves being as much as a foot long and half as wide. The second year the deep tap root produces a flowering stem varying from 2 to 6 feet tall. The yellow flowers are scattered on a dense spike, and each bilaterally symmetrical flower has five petals fused into a short tube. There are five stamens attached to the petals, with the upper three stamens covered with dense hairs. The pistil is prominent and somewhat bulbous at the top. Herbicides sprayed on these weeds often cause the stems to take on bizarre, twisted shapes.

Verbascum thapsus

105

One-flowered Helianthella

Helianthella uniflora
Sunflower Family

This species of *Helianthella* and its close relative *H. quenquenervis* can easily be confused with the common sunflower (*Helianthus annuus*), and possibly with the showy goldeneye (*Viguiera multiflora*). Foothills and lower canyon slopes are the most likely sites for this perennial which starts its blooming cycle immediately after the balsamroot ends its cycle. The leaves of *Helianthella* lack teeth, and flower heads are borne singly, being 1½ to 2½ inches in diameter. Like the cultivated sunflowers, they produce numerous squarish fruits bearing one seed. These seeds provide nourishment to a great many animals, from insects to birds to bears. In the past, sunflower seeds were eaten by Native Americans as food supplements.

Helianthella uniflora

Sulfur Buckwheat

Eriogonum umbellatum
Buckwheat Family

Throughout Utah, this highly variable species grows in a multitude of habitats, from sagebrush foothills to subalpine conifer communities. The miniature flowers appear in umbel-like clusters in midsummer or earlier, depending on elevation. Usually the flowering stalk is 10-14 inches tall, while numerous leaves grow very near to the ground and accumulate bits of organic matter which eventually becomes part of the soil. The generic name *Eriogonum* means "woolly knee or leg," and refers to the woolly stems of most species. Plants of this genus are difficult to distinguish. The tiny flowers have six petal-like parts which are held in five-toothed cups called involucres, and ovaries mature into three-sided fruits.

Eriogonum umbellatum

Calypso Orchid; Fairyslipper

Calypso bulbosa
Orchid Family

The Uinta Mountains is the most likely place to find this spectacular orchid, although it has been found in the Franklin Basin, near the Utah-Idaho boundary. Its presence indicates that the coniferous woods where the plants are found are relatively old and undisturbed, for the bulbous corm (stem) requires decaying wood and considerable organic matter. This shade-requiring orchid blooms early in the summer season. A single bilaterally symmetrical flower about 1¼ inch long hangs from a slender stalk 6 to 8 inches tall. The flowers resemble small lady's slippers with cup-like lips. Luer says the flowers are odorless and have no nectar. Pollination seems to be accomplished by means of insects which are visually deceived. In all the world there is only one species representing this genus.

Calypso bulbosa

Springbeauty

Claytonia lanceolata
Purslane Family

    Springbeauties are widespread throughout the counties of northern and central Utah, ranging from 4,500 to 10,000 feet. They are among the first flowers to follow the retreating snowbanks in the spring. Usually several stems grow from a tuberous underground corm. Each of these stems has two succulent leaves. Flower color varies from white to pink, and, in the whiter forms, pinkish veins add emphasis to the notched petals. There are five stamens and two sepals which remain long after the petals fall. The tuberous corms ($\frac{1}{2}$ to $\frac{3}{4}$ inches in diameter) were dug by Native Americans and eaten as we would eat potatoes. When the snow melts and the springbeauties, along with glacier lilies and buttercups, all blossom at the same time, a scene is presented that few will ever forget.

Claytonia lanceolata

113

Moss Campion

Silene acaulis
Pink Family

While this species is circumboreal in distribution, it is most likely to be seen in the Uinta Mountains of northeastern Utah. The small branches of this perennial plant form a tightly interwoven cushion connecting to a deep penetrating taproot. In the alpine ecosystem, such a growth habit raises the temperature inside the cushion and helps to create a microclimate which is more suitable for survival. Zwinger and Willard report that a moss campion plant may be 10 years old before flowering begins, since early growth energy goes into establishing a root system up to 4 or 5 feet deep that accumulates water and anchors the plant against almost constant wind. Each flower has five notched petals, five sepals, and ten delicate stamens. The leaves are short, narrow, and very close together, giving the cushion a mosslike appearance.

Silene acaulis

Sticky Geranium

Geranium viscosissimum
Geranium Family

Covered with sticky glandular hairs, this beautiful herbaceous perennial is found in sagebrush, grassland, and midmontane open woods, flowering from June through late August. It has strong branching stems from 15 to 30 inches tall and deeply lobed leaves. The rose-lavender flowers (1 to 1½ inches in diameter) are arranged on a plan of five, meaning that all parts are five or multiples of five. The capsule-like fruit, typical of the family, is beaked, due to the elongation of the style. As the capsule ripens, its longitudinal sections open with such recoiling force that the seeds are catapulted several feet outward from the parent plant. Note the prominent guide lines on the petals which help pollinators locate the nectar at the base of the petals.

Geranium viscosissimum

Prince's Pine; Pipsissewa

Chimaphila umbellata
Wintergreen Family

This circumboreal plant is a trailing and somewhat woody perennial, with leafy shoots and flowering branches. In northern Utah, it is found in the mid-montane, spruce-fir forest. The narrowly wedge-shaped leaves are evergreen and have marginal teeth which point forward. The flowers, borne in small umbellate clusters, are pink and distinctively bear ten stamens radiating around an unusually stout green ovary. If a hand lens is used to look at the stamens, the purple anthers reveal terminal pores through which pollen is shed, much like salt grains from a salt shaker. The name, pipsissewa, is evidently of Native American origin, and the plant was used for a variety of ailments such as rheumatism and fevers. The flowers appear in late July or early August.

Chimaphila umbellata

Twinflower

Linnaea borealis
Honeysuckle Family

This sweet-scented, trailing evergreen is found in the coniferous forest of the Uinta Mountains and is easily overlooked. From the prostrate branches, flower stalks rise which bear two pendant pink flowers which are less than ½ inch long. The funnel-shaped corolla is almost equally five-lobed, but there are only four stamens. The tiny dry fruit which follows the flower has hooded bristles that readily become attached to the fur and feathers of animals. The Swedish botanist, Linnaeus, who established the binomial system of nomenclature, considered this his favorite flower and, hence, named it after himself. The species is circumboreal and blossoms in July and August. The Kootenay Indians reportedly consumed a tea made from leaves of the twinflower.

Linnaea borealis

Globemallow; Wild Hollyhock

Iliamna rivularis
Mallow Family

Along streams and roadsides from the foothills up to 8,000 feet in canyons, the gorgeous pink flowers of globemallow add their distinctive color during July and August. The stems are stout, branched, and reach a height of 4 to 5 feet. The maple-like leaves are 2 to 8 inches across, generally with five lobes. The individual flowers are up to 2 inches in diameter. The filaments of the stamens are united into a tube surrounding the style. Irritating hairs cover the fruits, which, upon drying, break open like segments of oranges. In some areas of the Intermountain West, this is one of the first herbaceous species to appear after a fire. The seeds require some heat from fires or scouring before they can germinate.

Iliamna rivularis

Fireweed

Epilobium angustifolium
Evening Primrose Family

From foothills to subalpine, this circumboreal species occupies many habitats. The common name refers to its ability to populate burned-over and logged areas with a beautiful cover of pink to magenta flowers. The reddish stems, bearing many alternate lance-like leaves, may reach a height of 6 feet. The slightly irregular flowers are built on a plan of four. Note particularly the four spreading stigma lobes which rise above all the flower parts. In the fall, the slender inflorescence takes on a fluffy white appearance because the tips of the seeds are covered with long white hairs. The forests in this area are showered annually with the airborne seeds of this plant. Successful invasion by the fireweed, however, depends on the reduced plant competition of disturbed sites. Young shoots, young leaves, and flower bud clusters are abundant and edible, especially as potherbs.

Epilobium angustifolium

Elephanthead

Pedicularis groenlandica
Figwort Family

A close examination of a single flower will reveal why this plant has such a descriptive common name. The upper lip of the corolla has a long upward curving beak which encloses the style. Two petals of the lower lip are shaped like ears. Together, the parts of this irregular flower have an amazing resemblance to an elephant's head. The height of the plant varies from 10 to 20 inches, and the leaves are all pinnately divided and quite fern-like. Look for the pink to magenta racemes in wet meadows and along streams from 7,000-10,000 feet during July and August. Pollination of individual flowers is brought about by a complex vibrating action of visiting bees. The name *Pedicularis* is derived from Latin meaning "louse," and was based on the old belief that cattle grazing on one species in Europe would become covered with lice.

Pedicularis groenlandica

127

Carpet Phlox

Phlox hoodii
Phlox Family

Phlox is from the Greek meaning "flame," and many species of the genus have been brought into cultivation, especially for rock garden use. This species occurs in open sites in sagebrush communities from the valleys to the midmontane zone. The stiff and spiny leaves are covered with long or cobwebby hairs. The tube of the corolla is about $\frac{1}{2}$ inch long and is crowned with five lobes which extend at right angles. Petal color can be variable, but is usually some shade of pink. Longleaf phlox (*Phlox longifolia*) will occur in the same habitat, but it is easily recognized because it is taller than the carpet phlox, by up to 6 inches, and has softer, longer leaves which reach 1 inch in length. Both species flower in May and early June.

Phlox hoodii

6 - 2 - 12
Grand Canyon
White

129

Lewis Monkeyflower

Mimulus lewisii
Figwort Family

Hiking Wasatch and Uinta mountain trails from 7,000 to 10,000 feet, one is very likely to see the Lewis monkeyflower growing close to a small stream or on a wet ledge. The plant was named after Captain Lewis of the Lewis and Clark expedition (1804-1806), who found the plant near Glacier National Park. The five petals are united into a tube with spreading corolla lobes (three down and two up). The lower lobes have yellow ridges with conspicuous hairs. Such a structure is well adapted to specialized pollinators such as flies, bees, and hummingbirds. When a bee crawls into the wide opening of the corolla for nectar, its back is dusted with pollen which it then carries to the next flower. This tall perennial grows in clumps from underground rhizomes. The upright stems bear sessile, opposite leaves and may reach a height of $2\frac{1}{2}$ feet.

Mimulus lewisii

131

Scarlet Gilia; Skyrocket

Ipomopsis aggregata
Phlox Family

This plant adds striking color to the sagebrush community during June and July, but can also be found in the midmontane zone during August and early September. Usually this species is biennial, producing only a small clump of basal leaves the first year, followed by $1\frac{1}{2}$ to $2\frac{1}{2}$ foot flowering stems the second year. The flaring corolla lobes are bright red with a yellowish mottling on the inside. This color arrangement is especially attractive to hummingbirds that thrust their bills down the tube of the corolla seeking nectar at the base. In the hovering and collecting process, the bird's head becomes covered with pollen, and when it hovers at the next flower, pollination is assured. In some localities of Utah, a much lighter color phase will be found, and these corollas are attractive to hawk moths.

Ipomopsis aggregata

133

Firechalice

Epilobium canum
Evening Primrose Family

Few other wildflowers can bloom as long as this species. Starting in late July, the red tubular flowers continue into late October. This somewhat woody perennial can be erect to decumbent, and can occupy an elevational range from 4,500 to 9,000 feet, usually on rocky cliffs and ledges. The scarlet flower tube (hypanthium) is about 1 inch long and trumpet shaped. Four sepals and four two-lobed petals are inserted at its top. A long slender style bearing the four-lobed stigma protrudes beyond the petals. The flowers, with their red color, tubular form, and abundant nectar, attract hummingbirds, especially in late summer. The leaves are usually hairy and have a many-toothed margin.

Epilobium canum

135

Asian Poppy

Roemeria refracta (Papaver rhoeas)
Poppy Family

This alien species, which is native to western Asia, puts on a splash of red color in Box Elder and Cache counties during late May and early June, in a display of such magnitude that one must see it to believe it. Sometime in the decades of the thirties or forties, this annual became established in the grain fields of northern Utah, and since then seems to have become a permanent resident. The flowers have two sepals and four petals, and each petal has a black spot at the base. Like other poppies, the Asian poppy has numerous stamens. The 5 to 18 carpels are united and eventually form a conspicuous capsule with seed dispersal pores at the apex. Medicinally, the Asian poppy has been used as a sedative and an analgesic, but, unlike the opium poppy, it is not a source of narcotics.

Roemeria refracta (Papaver rhoeas)

Eaton Penstemon

Penstemon eatonii Gray
Figwort Family

This spectacular penstemon, also known as fire-cracker penstemon, is adapted to sagebrush, mountain brush, and aspen communities. The bilaterally symmetrical corollas hang downward slightly in long open clusters above sparsely leafy stems which reach up to 30 inches tall. The leaves are entire and the upper ones are sessile, clasping, and often heart shaped. More than 60 species of *Penstemon* survive in Utah, and at least four species have scarlet flowers. Only the eaton penstemon has a corolla tube that does not conspicuously flare open at the mouth. Like other penstemons, there are four anther bearing stamens and one sterile stamen called a staminode. Nectar is secreted at the base of the corolla and attracts hummingbirds for the pollination process. With an elevational range from 4,000 to 9,000 feet, the plants can flower from May through early August.

Penstemon eatonii Gray

139

Wavy-leaved Paintbrush

Castilleja applegatei
Figwort Family

Most people recognize the paintbrush flower, but few care to separate the 14 species of this genus in Utah. The difficulty of identification arises from hybridization and other genetic problems such as polyploidy. The wavy-leaved paintbrush can be found in dry to mesic sites from the valleys to the subalpine during the months of June through August. The actual flowers of the eye-catching paintbrush group are narrow, tubular, and partly greenish-yellow. The vivid scarlet of the leafy bracts surrounding the flower provide the color for this species complex. The calyx generally has four lobes, and the corolla has a narrow folded upper lip and a lower lip which may be three lobed. Most species of this genus are partial parasites on such plants as sagebrushes and grasses. The root connections established on the host provide nutrients and water. For this reason, the paintbrush species usually cannot be cultivated.

Castilleja applegatei

6-2-12
Grand Canyon

Tapertip Onion

Allium acuminatum
Lily Family

There are at least ten species of wild onions in Utah and about 500 species of *Allium* in the world. These plants have been used for their edible bulbs since ancient times. Many animals, such as bear and elk, also use these odoriferous plants. All alliums have the same distinctive flower structure—three sepals, three petals, six stamens and three fused carpels. The flowering stem of the tapertip onion is 6 to 12 inches tall. The petal tips are tapered and spreading. The basal leaves are grasslike and usually wither by flowering time in June or July. Look for them in dry sites on the foothills and at mid-elevations in the mountains. The generic name is from the Latin word *allium* which means "garlic."

Allium acuminatum

Few-flowered Shooting Star

Dodecatheon pulchellum
Primrose Family

The shooting star species of Utah are similar to the cultivated cyclamen, having flowers in an umbel and petals which are reflexed backward. Some flower enthusiasts have described the umbels as miniature firework displays. The flowering stems grow to 12 to 15 inches tall and the flowers are from $\frac{1}{2}$ to 1 inch long. There are five sepals and five petals which bend 180 degrees away from the colorful stamens. The stamens form a dark colored cone around the style. Pollination studies reveal that these flowers are uniquely adapted to "buzz" visits by bumblebees. These bees grasp the base of the staminal tube and vibrate the flower, thus shaking the pollen grains onto the ventral surface of the insect. When the next flower visit occurs, the elongated style and stigma receive the essential pollen. The species occupies meadows and streamsides from high valleys to subalpine, and flowering occurs in May and July.

Dodecatheon pulchellum

Utah Sweet Pea

Lathyrus pauciflorus
Pea Family

Even though this species is never seen in large numbers, it is frequently found in moist areas in the foothills and canyons climbing over sagebrush and other shrubs of the midmontane zone. Some terminal leaflets of the pinnately compound leaves are modified into tendrils. The presence of these twining tendrils, along with bilaterally symmetrical flowers, easily sets this plant apart from other legumes of the area. This weak-stemmed perennial gives rise to only one or two flowering stalks per plant. Each raceme produces only four to eight flowers, which is the reason for the specific epithet, *pauciflorus*, meaning "few flowers." Several taxa in this genus vary greatly in stipules and leaflets and, hence, are sometimes difficult to distinguish. Flowering is best during June and July.

Lathyrus pauciflorus

147

Mountain Lover

Paxistima myrsinites
Stafftree Family

This low evergreen shrub has the appearance of a boxwood and grows in a wide range of environments from the foothills to the subalpine, usually under the shade of a variety of woody species. The leaves are opposite and leathery and have sharp teeth along the margins. The diminutive flowers are red to maroon, are constructed on a plan of four, and are positioned in the axils of sessile leaves. The fleshy petals alternate with the four short stamens. If a hand lens is used on the tiny blooms, one can appreciate the variety of size and diversity that evolution has produced in flowering plants. Flowering in northern and central Utah occurs in late May and early June.

Paxistima myrsinites

Steershead

Dicentra uniflora
Fumitory Family

Growing in gravelly soil, this species is usually found in mountain brush and spruce-fir communities up to 9,000 feet. This perennial herb is one of the most beautiful harbingers of spring. The steershead is unique in its flower structure. Like other bleeding hearts in the family, it has two sepals, four petals and six stamens. Only 2 to 4 inches tall, the plant has a single flower at the tip of each leafless stem. The longer outer petals are curved backward, exposing the tops of the inner petals. The whole flower is about $\frac{3}{4}$ inch long, and it is easy to see the connection to the common name. The long petioled leaves are delicate and subdivided into many oblanceolate segments. It flowers within a few days after the snow melts, which means April to June, depending on elevation.

Dicentra uniflora

Parry Primrose

Primula parryi
Primrose Family

Because of its clustered magenta flowers, this large alpine herbaceous plant stands out conspicuously along streams and below melting snowbanks. Close examination of a flower reveals that the five petal lobes are joined at their base into a narrow tube. These flowers have the odor of carrion and, undoubtedly, attract muscoid flies to accomplish pollination. The leaves are basal and somewhat succulent. June and July are the best months for flowering. A related species, *P. maguirei* (maguire primrose), is found only in Logan Canyon on limestone cliffs. This federally listed threatened and endangered (proposed) species is much smaller (7 to 10 inches tall) than the parry primrose, and has fewer flowers.

Primula parryi

Primula maguirei

153

Northern Sweetvetch

Hedysarum boreale
Pea Family

Occupying dry, open habitats, the northern sweetvetch can be found from sagebrush foothills to mountain brush in the canyons. Since the flowers of this species closely resemble those of milkvetches and locoweeds, it is prudent to look for fruit to verify identification. Sweetvetch pods are flattened and are remarkably constricted between the seeds so that each segment appears round. The other mentioned plants have typical fruits shaped like garden peas. Leaves are pinnately compound with many leaflets. The keel of the flower is nearly straight and longer than the wings. June and July are the month for flowering. Historically, Native Americans have collected and eaten the non-poisonous roots, which have a licorice-like taste.

Hedysarum boreale

155

Western Columbine

Aquilegia formosa
Buttercup Family

Restricted to the midmontane and riparian habitats of Box Elder and Tooele counties, this columbine is worth searching for, especially along the Utah-Idaho border. The five petals called "spurs" are bicolored—red at the nectar holding end and yellow at the flaring end. Five red sepals alternate with the petals. Note how the flowers usually hang down. This is in contrast to other species which are straight out in position. This combination of red perianth member and drooping flowers favors the visitation of hummingbirds which can hover in their visits. Long-tongued moths and butterflies may also pollinate, but bees do not see red and, hence, are not seen at the pollination time of midsummer. Because few barriers exist to prevent gene exchange, plant breeders have produced multi-colored hybrids for the garden trade. Even so, seeing this native wildflower in its natural setting is a joy to remember.

Aquilegia formosa

Pink Wintergreen; Shinleaf

Pyrola asarifolia
Wintergreen Family

The nodding flowers are light pink to rosy-red and are arranged in slender racemes on stems 8 to 15 inches tall. There are five petals and ten stamens, the anthers of which release their pollen through terminal pores. The style is characteristically bent to one side and often has a ring or collar below the stigma. The ovate leaves are basal, relatively thick, and shiny. Look for this evergreen perennial in wet sites in spruce-fir forests, especially in the Uinta and Wasatch mountains up to 9,000 feet. The fruit is a dry capsule. At least four other species of pyrolas occur in similar habitats of northern Utah. Flowering extends from June through August.

Pyrola asarifolia

Prairiesmoke

Geum triflorum
Rose Family

Prairiesmoke's nodding flowers and the fruiting heads of its feathery styles have been responsible for many common names, such as old man's whiskers, China bells and long-plumed avens. The flowers usually arise in threes, and this explains the specific epithet. The stems are 7 to 18 inches tall and have mostly basal pinnately compound leaves. The whole plant is covered with soft hairs, and this feature makes it challenging as a photographic subject. The pinkish sepals are fused at the base to form a bowl-shaped cup which partially hides the petals and numerous stamens. After fertilization, the flowers become erect and the many styles elongate so that the wind will eventually disperse the seed. Look for the flowers in sagebrush and mountain brush of the midmontane zone in July and August.

Geum triflorum

Western Monkshood

Aconitum columbianum
Buttercup Family

Monkshoods inhabit wet meadows and stream banks up to 9,000 feet in the major canyons. The stem is stout and varies from 2 to 6 feet in height. The upper sepal of this bilaterally symmetrical flower is modified into a hood-shaped structure and is responsible for the common name. In addition, there are two broad sepals at the side and two small ones below. Concealed within the hood are two petals and numerous stamens. The flowers are generally purple, but albinos occasionally occur. This species depends upon bumblebees for pollination; smaller insects haven't the strength to push aside floral parts to reach the nectar. Like the larkspur, the plant contains poisonous alkaloids, especially in the roots. It flowers from June through September.

Aconitum columbianum

163

Roundleaf Harebell

Campanula rotundifolia
Bellflower Family

The generic name of this perennial means "little bell," and the specific epithet, *rotundifolia*, refers to the roundish, heart-shaped, basal leaves. While these basal leaves wither early, the narrow leaves of the stems remain. A conspicuous feature of the flowers is that although the buds grow erect, the open flowers droop or are horizontal—their position protects the pollen from the rain. Occasionally white blossoms will grace the stems of this circumboreal species. Throughout July and August this delicate herb is frequently seen in open coniferous woods of the canyons and the subalpine. This plant spreads rapidly in rock gardens by rhizomes.

Campanula rotundifolia

Utah Milkvetch

Astragalus utahensis
Pea Family

In the sagebrush community of old lake terraces and foothills, it is relative easy to find the Utah milkvetch in full bloom as early as late April. Also called locoweed, the genus *Astragalus* contains many species which are similar and difficult to distinguish. This species is quite distinctive, however, with heavy pubescence covering both surfaces of the pinnately compound leaves. The fruit pods, called legumes, really set this plant apart. They are covered by long woolly hairs that prevent the destruction of seeds by seed-eating insects. Locoweeds contain alkaloid-like substances that can cause serious loss of livestock. These plants do not present a threat to man since they are seldom eaten by humans. One can enjoy the purple to pink color of their butterfly-like blossoms.

Astragalus utahensis

Common Camas

Camassia quamash
Lily Family

This member of the lily family is an onion-like plant arising from a bulb which has been used by Native American tribes as an important food. The names "camas" and "quamash" are Indian names that have been combined in the scientific name. Camas plants inhabit wet meadows or low, non-salty areas that remain quite wet during May or June. A flowering stalk reaches a height of 18 to 24 inches and is crowned by an elongating inflorescence of purplish-blue flowers, 1 to 1½ inches in diameter. Harrington says the bulb seems to be lacking in starch, although its sugar content is high. The bulbs were dug during any season, but it was safest when its purple flowers distinguished it from the highly poisonous death camas, which has much smaller white flowers.

Camassia quamash

Sugarbowl; Hairy Clematis

Clematis hirsutissima
Buttercup Family

This extraordinary herbaceous plant reaches a height of 1 to 2 feet and inhabits sagebrush to spruce-fir communities from 6,000 to 8,500 feet. The flower has no petals and the four purple, petaloid sepals are somewhat obscured on the outer surface by a covering of cobwebby hairs. Many stamens and pistils are enclosed in the perianth bowl. The fruits are one seeded structures called achenes, and they have long feathery styles. The leaves are opposite and are two to four times pinnately divided. Because of the thickness and wrinkled texture of the sepals, the plant has been called leather flower. The specific epithet *hirsutissima* means "very hairy." Unlike most other *Clematis* species, it is not a vine. Blooming occurs in early June through July.

Clematis hirsutissima

171

Whipple Penstemon

Penstemon whippleanus
Figwort Family

Within the montane zone, this distinctive species (one of at least 25 penstemon species) appears on wooded slopes in the subalpine and continues upward into the alpine zone. The flowers of this penstemon come in two color phases—deep wine lavender and the less common creme with purple veins. As in many other penstemons, it has multiple stems, opposite leaves, and a corolla with five joined petals. A hand lens will reveal the numerous glandular hairs that cover the sepals and petals. There are four stamens with anthers and a fifth stamen without an anther but usually with a cluster of hairs at its tip. Such specialized flower structure is typical of the figwort family and indicates a long evolutionary history of adaptation to rather specific pollinating agents. July and August are the blooming months.

Penstemon whippleanus

Beckwith Violet

Viola beckwithii
Violet Family

There are at least seven or eight naturally occurring species of violets in northern and central Utah. The Beckwith violet is the one which draws the most attention during April and May. Blue violets typically inhabit cool, moist sites in the canyons. This species however, grows only in poorly developed soils on the foothills. By late June, the plants have dried up and appear dead. The two upper petals are purplish while the three lower ones are white streaked with purple nectar guides against yellow bases. The leaves are quite distinctive, somewhat fleshy, and divided into a number of elongated leaflets. It is likely that all species of violets are edible, but this Beckwith violet should be left in place as foothill habitats in northern Utah are under pressure from encroaching urban sprawl.

Viola beckwithii

175

Mountain Bluebell

Mertensia ciliata
Borage Family

Reaching a height of 3 feet, this showy bluebell inhabits subalpine and alpine areas in meadows and especially along streambanks. The lanceolate alternate leaves and succulent stems are enjoyed by many animals of the forests. The tubular flowers are purplish in bud but rapidly turn blue as blossoms expand to full size. The five fused petals that form the bell have a lower tube and a flaring upper section. Five stamens are attached to the inside of the corolla. The peak blooming time is late July and early August. Two other bluebell species are much shorter (less than 15 inches) and flower in the foothill communities during April and May (*M. brevistyla*, short-styled bluebell, and *M. oblongifolia*, sagebrush bluebell).

Mertensia ciliata

Silvery Lupine

Lupinus argenteus
Pea Family

Lupines as a group are easily recognized by typical pea-like flowers, hairy fruit pods, and the palmately compound leaf with five to several leaflets at the top of a long petiole. The plants grow in dense colorful clumps in the foothills and the open forests of the canyons. The genus is taxonomically very difficult. As a result, botanists have described over 600 species. The problem within the group is that many species hybridize, yielding a broad spectrum of intergrading forms. Lupines benefit the soil because their root nodules contain nitrogen-fixing bacteria. However, lupines contain lupinine and other related toxic alkaloids which may have serious consequences for livestock. Flowering begins in June and continues into mid-August.

Lupinus argenteus

Showy Daisy; Showy Fleabane

Erigeron speciosus
Sunflower Family

With over 25 species of *Erigeron* in Utah alone, one can imagine the difficulty in distinguishing them. The specific epithet *speciosus* means "showy," and well describes the brightly colored heads of this winsome species. Showy fleabane stems may be up to 25 inches high and bear several flower heads, each with from 70 to 100 slender purplish ray flowers. The tubular disk flowers are yellow orange. Separating species of *Erigeron* from *Aster* can also be frustrating. Generally, many fleabanes flower earlier in the summer season than do asters. Also, the ray flowers of *Erigeron* are more numerous and narrower. The plants are found in mesic sites of canyons and extend up into the subalpine. Flower heads are 1½ to 2 inches in diameter and begin attracting bees in June.

Erigeron speciosus

Fringed Gentian

Gentianopsis detonsa
Gentian Family

The gentians as a group are cosmopolitan and include nearly 1,000 species, especially in the arctic and mountains. Fringed gentian is an annual of circumboreal distribution. Wet meadows and streamsides of high valleys and the midmontane zone during July and August are the best locations to seek the bell-shaped flowers. The four petals are fused into a corolla about two inches long, the lobes of which are delicately fringed. The stems, between 6 and 15 inches high, are often in a cluster, with several pairs of leaves. The National Park Service selected this plant to be the official flower of Yellowstone National Park.

Gentianopsis detonsa

183

Mountain Bog Gentian

Gentiana calycosa
Gentian Family

As the common name implies, the range of this subalpine and alpine plant is limited to between 7,000 to 10,000 feet, and it is found primarily along streambanks. The stems vary from 6 to 13 inches high and possess three-veined leaves with smooth margins. Usually there is a single flower about 1½ inches long, the interior of which is spotted with grayish-green dots. Note also that between the five major corolla lobes are short, double-pronged projections. The specific epithet *calycosa* refers to the cup-shaped calyx of five united sepals. This species is a perennial and has thick, fleshy roots.

Gentiana calycosa

Sky Pilot

Polemonium viscosum
Phlox Family

Sky pilot is truly a symbol of the lofty alpine zone, especially in the Uinta Mountains where it will occupy boulder fields and scree slopes. The clusters of funnel- or bell-shaped flowers sparkle with various shades of blue to blue-violet, and the bright orange pollen is often seen on the legs of the bees which systematically forage from plant to plant. This and other *Polemonium* species have one striking feature in common—a fetid skunky odor. This unpleasant aroma is traceable to the sticky glandular hairs that cover stems, leaves, and sepals. The leaves are up to 6 or 7 inches long and pinnately compound. Each leaflet is divided into 3-5 lobes. Flowering occurs in July and August. Two taller dwelling species *P. occidentale* and *P. foliosissimum* grow in wet meadows of the midmontane zone.

Polemonium viscosum

Wasatch Beardtongue

Penstemon cyananthus
Figwort Family

Paralleling the Wasatch Mountains, this species of penstemon often fills disturbed sites with handsome patches of blue from May through July. The several stems with ovate to lanceolate leaves may reach a height of 3 feet. The common name, beardtongue, refers to the fact that only four of the stamens produce pollen. The fifth is sterile, flattened, tongue-like, and bearded at the terminal end with golden hairs. The bilaterally symmetrical corolla varies from $\frac{3}{4}$ to $1\frac{1}{4}$ inches long, and its shape and swollen base favors pollination by bumblebees. At least 25 species of penstemons are scattered across northern Utah, and the color variation is truly amazing—red, violet, blue, white, cream, and many others. Numerous small seeds are concentrated in dehiscing capsules.

Penstemon cyananthus

189

Wild Blue Flax; Lewis Flax

Linum perenne
Flax Family

Some botanists consider this species to be a sub-species, *lewisii*. It is named in honor of Lewis of the Lewis and Clark expedition. The plants inhabit dry rocky soils of the valleys and extend up to the subalpine ridges. The numerous flowers are borne on slender stems of from 1 to 2 feet tall which bend and bow to every passing breeze. The five blue petals (most color films do not record this hue correctly) are extremely fragile and will fall at the slightest handling. The five styles are longer than the five stamens. The blossoms open early in the morning and usually close late in the afternoon. The brown seed capsules, about $\frac{1}{4}$ inch across, contain numerous seeds that are rich in oil. Native Americans of the Pacific Northwest used the plant to make their thread and fishing tackle. The plant blooms from June through August.

Linum perenne

191

Thickstem Aster

Aster integrifolius
Sunflower Family

Some 16 species of *Aster* range from the valleys to the alpine and manage to inhabit many moisture levels. While many asters are difficult to distingish, the thickstem aster has some distinctive features that set it apart. The rather tall stems (10 to 24 inches) have only a few flower heads in a narrow inflorescence. The sparse, purple rays vary from 10 to 25 and surround the yellow-orange disk flowers. The involucral bracts below the flowers and flower stalks are covered with glandular hairs. Late August and early September are less drab because of this ragged but beautiful perennial. This particular aster will germinate in disturbed sites as well as dry meadows and open forests.

Aster integrifolius

193

Western Clematis

Clematis occidentalis
Buttercup Family

This woody-stemmed clematis is one of the few climbing vines native to northern Utah, and it has many features to set it apart from the other woody plants. The leaf stalks or petioles act like tendrils and attach themselves to large shrubs and small trees. The flowers lack true petals, but are conspicuous because of the blue to lavender petal-like sepals and numerous stamens. The four sepals are lanceolate and 1½ to 2 inches long and are attractively accented by darker veins. The fruits which develop from numerous pistils have long fuzzy styles. The feathery styles aid in scattering the one-seeded achenes. The generic name *Clematis* is derived from the Greek "klemma," meaning a vine-branch. This species is most apt to be seen in midelevations with spruce-fir and mountain brush communities, and it starts blooming in May.

Clematis occidentalis

Ballhead Waterleaf

Hydrophyllum capitatum
Waterleaf Family

The first visitors of each travel season will find this handsome plant along with violets and springbeauties during May and June, especially among the maples and aspen of the canyons. The specific epithet *capitatum* means a "head," and refers to the flower cluster. *Hydrophyllum* is Greek meaning "waterleaf," and is named after the watery or succulent leaves often found in the family. The light violet to purple flowers (about ⅜ inches long) are in a dense, ball-like cluster which has a fringed appearance because the stamens and bilobed stigmas are held conspicuously above the tubular corolla. The stems and leaves are somewhat succulent, and the latter are pinnately cleft and fernlike. The tender shoots and roots were used as potherbs by Indians and settlers.

Hydrophyllum capitatum

Low Larkspur

Delphinium nuttallianum
Buttercup Family

This species is widely distributed from foothills to subalpine and from sagebrush to pinyon-juniper, with peak flowering in May and June. The plant commonly stands less than 2 feet high, and the stem develops from clusters of tuberous roots. The upper sepal of the flower extends backward into a prominent spur. The spur sepal plus four other sepals expand to reveal four smaller petals of lighter color with conspicuous purple veins. Hummingbirds are frequent visitors seeking the nectar at the base of the spur. The leaves are palmately lobed or divided. Although it has practically no effect on naturally occurring wildlife, the low larkspur contains a combination of alkaloids which are toxic to livestock. The showy larkspur, *D. occidentale,* is tall (3 to 6 feet), has numerous flowers in an elongated raceme, and occupies moist sites in the canyons. It, too, is poisonous to livestock.

Delphinium nuttallianum

Woodland Pinedrops

Pterospora andromedea
Wintergreen Family

The reddish-brown stems may reach a height of 3 feet and are covered with sticky hairs. The non-green leaves are small and scale-like; they are restricted to the lower half of the stem. Careful research has shown that this plant is a mycotroph, which means that it lives as a parasite on soil fungi. The fungi, in turn, have mycelial connections to the roots of forest trees. Thus the fungus permits the woodland pinedrop to parasitize the trees indirectly. The urn-shaped nodding flowers have a five-parted calyx and a united corolla. The fruits of the inflorescence mature into brown capsules that release great quantities of minute winged seeds. Stems grow for only one year, but the dried stems may last into the following year. Some Native Americans used an infusion of this plant to prevent nosebleeding. The Uinta Mountains are most apt to have this unique species.

Pterospora andromedea

Clustered Broomrape

Orobanche fasciculata
Broomrape Family

     A flower aficionado should experience a thrill upon finding this small and delicate parasite (up to 6 inches high). It has been reported on a variety of host plants, but sagebrush plants seem to be the most frequent host. Usually there is a thick underground stem, the specialized roots (haustoria) of which penetrate the host's roots to withdraw water and nutritional compounds. The calyx and tubular corolla are brownish to yellowish with some purple color within the tube. There are two pairs of stamens, and the overall appearance is that of a miniature *Penstemon* species. Parasitism as a way of life is often looked upon with disdain, but it is actually a highly successful strategy for survival that has evolved many times in a number of plant families. Look for flowers in June through early August.

Orobanche fasciculata

203

Striped Coralroot

Corallorhiza striata
Orchid Family

Most coralroots are devoid of the green pigment, chlorophyll, and cannot manufacture their own food. They are completely dependent on a group of saprophytic fungi which survive in the duff of the coniferous-aspen forests. Thus, the coralroots live as parasites on the mycelial filaments of the fungus. This survival strategy is called mycotrophy (literally meaning fungus nourishment). The broad lower petal (lip) is almost completely purple, and the upper sepals and petals are brownish overall with 3 red to purple stripes. The leaves are reduced to sheaths on the lower part of the stem. Even though the generic name means "coral root," there is no root per se but rather a hard mass of rhizome tissue associated with a fungus. The rhizome may remain dormant for years after producing flowering stems. This coralroot produces flowers in May and June.

Corallorhiza striata

Western Coneflower

Rudbeckia occidentalis
Sunflower Family

August is the month to watch for this stout perennial which varies from between 3 to 6 feet tall. The distinctive flowering heads are on tall leafy stems. The colorful ray flowers seen in other composites are lacking in this species, and yet the numerous disk flowers are obviously displayed on a cylindrical cone nearly 2 inches long. In spite of the diminutive size of the disk flowers, they provide considerable nectar and pollen for frequent visits of bumblebees. The green, involucral bracts at the base of the cone are in two to three series and mostly unequal in length. This species grows in meadows of the canyons but is especially common in sites such as trails or areas overgrazed by cattle and sheep.

Rudbeckia occidentalis

207

Leopard Lily; Purplespot Fritillaria

Fritillaria atropurpurea
Lily Family

Leopard lilies are easily missed if the trail hiker in the Wasatch and Uinta mountains isn't watching for them. The fast growing stem (up to 20 inches tall) bears two to four nodding flowers. The stems and narrow leaves are derived from a flattened bulb with many bulblets for asexual reproduction. When one of the flowers is examined closely, the yellowish splotches are revealed against a brown or purplish background. There are six perianth members of equal length and six very conspicuous yellow anthers. The odor emanating from the flowers, especially in the morning, is very unpleasant to humans but very attractive to flies, which are probably responsible for pollination. The plants grow in open forests or grassy slopes of the canyons. It generally flowers in June.

Fritillaria atropurpurea

Helleborine

Epipactis gigantea
Orchid Family

Fourteen of the twenty-one northern and central Utah counties have reports of this widespread orchid. The stems are erect and stout and bear a goodly number of clasping, ovate leaves. The flowers are large, up to an inch across, and with the use of a hand lens they become fascinating as an example of specialization to a specific insect pollinator. The brownish to purple perianth members attract hover flies. While these flies mimic bees in appearance, they are true flies that pick up pollinia (clusters of pollen) and transfer them to other orchid flowers. These perennial orchids arise from creeping rhizomes which may be transplanted into gardens with some success. Normally this helleborine will be found in shady areas along rivers, streams and seeps. The blooming period extends from June through July.

Epipactis gigantea

SELECTED REFERENCES

Albee, B. J., et al. 1988. *Atlas of the Vascular Plants of Utah.* Utah Museum of Natural History, Salt Lake City, Utah.

Coffey, T. 1993. *The History and Folklore of North American Wildflowers.* Facts on File Inc., New York.

Elias, T. S. and P. Dykeman. 1982. *Field Guide to North American Edible Wild Plants.* Van Nostrand Reinhold Company, New York.

Harrington, H. D. 1967. *Edible Native Plants of the Rocky Mountains.* University of New Mexico, Albuquerque, New Mexico.

Luer, C. A. 1975. *The Native Orchids of the U. S. and Canada.* New York Botanical Garden, New York.

Schmutz, E. M. 1979. *Plants that Poison.* Northland Press, Flagstaff, Arizona.

Shaw, R. J. 1989. *Vascular Plants of Northern Utah: An Identification Manual.* Utah State University Press, Logan, Utah.

Shaw, R. J. 1992. *Wildflowers of Grand Teton and Yellowstone National Parks.* Wheelwright Press, Salt Lake City, Utah.

Weiner, M. A. 1972. *Earth Medicine—Earth Foods.* Collier Books, New York.

Welsh, S. L. et al. 1993. *A Utah Flora.* 2nd edition, revised. Print Services, Brigham Young University, Provo, Utah.

Whitson, T. D. et al. 1991. *Weeds of the West.* The Western Society of Weed Science, University of Wyoming, Laramie, Wyoming.

Zwinger, A. H. and B. E. Willard. 1972. *Land Above the Trees.* Harper and Row, New York.

PARTS OF A FLOWER

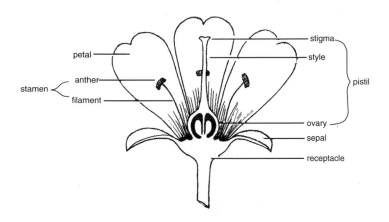

Common Botanical Terms

Anther. The pollen-bearing part of a stamen.

Biennial. A plant which produces seed during its second year of life and then dies.

Bract. A modified leaf associated with the flower.

Calyx. The outer set of perianth segments, usually green; a collective name for sepals.

Capsule. A type of dry seed pod producing many seeds.

Carpel. A single "inrolled spore-bearing leaf" (stigma, style, and ovary); a simple pistil or one of the units forming a compound pistil. A good example is the pod of a pea or a peanut.

Corm. A short, bulb-like, underground stem.

Corolla. A collective term referring to the petals of a flower.

Dehiscent. Opening spontaneously when ripe to discharge the contents.

Disk flower. A central flower of a composite inflorescence (such as the center of a sunflower).

Inflorescence. A flower bearing branch or system of branches.

Node. A point on a stem from which a leaf and bud arise.

Pedicel. Stalk of a single flower or a grass spikelet.

Perennial. A plant which lives for more than two years.

Perianth. The collective name for the sepals and petals.

Petiole. The stalk of a leaf.

Pinnate. A compound leaf with the leaflets on two opposite sides of an elongated axis.

Ray flower. A marginal flower of a composite inflorescence.

Sepal. One of the parts of the outer whorl of floral leaves, usually green in color.

Spike. An elongated flower cluster, having the individual flowers very close to the stem.

Spur. A tubular or saclike projection from a blossom.

Stamen. A male organ of a flower which produces pollen.

Stigma. The top part of a pistil which receives the pollen.

Succulent. Juicy or pulpy.

Umbel. An inflorescence in which all the pedicels arise from the same point on the stem.

INDEX